Wolfgang Herschel

On the Remarkable Appearances at the Polar Regions of the Planet Mars, the Inclination of Its Axis, the Position of Its Poles, and Its Spheroidical Figure...

Wolfgang Herschel

On the Remarkable Appearances at the Polar Regions of the Planet Mars, the Inclination of Its Axis, the Position of Its Poles, and Its Spheroidical Figure...

ISBN/EAN: 9783337213466

Printed in Europe, USA, Canada, Australia, Japan

Cover: Foto ©berggeist007 / pixelio.de

More available books at **www.hansebooks.com**

PHILOSOPHICAL

TRANSACTIONS.

XIX. *On the remarkable Appearances at the Polar Regions of the Planet* Mars, *the Inclination of its Axis, the Position of its Poles, and its spheroidical Figure; with a few Hints relating to its real Diameter and Atmosphere.* *By* William Herſchel, *Eſq. F. R. S.*

Read March 11, 1784.

WHAT I have to offer on the ſubject of the remarkable appearances at the polar regions of Mars, as well as what relates to the inclination of the axis, the poſition of the poles, and the ſpheroidical figure of that planet, is founded on a ſeries of obſervations which I ſhall deliver in this paper; and

Vol. LXXIV. I i after

after they have been given in the order they were made, it
will be eafy to fhew, by a few deductions from them, that my
theory of this planet is fupported by facts which will fufficiently authorife the conclufions I have drawn from them. For
the fake of better order and perfpicuity, however, I fhall treat
each fubject apart, and begin with the remarkable appearances
about the polar regions. The obfervations on them were made
with a view to the fituation and inclination of the axis of
Mars; for to determine thefe we cannot conveniently ufe the
fpots on its furface, in the manner which is practifed on the
fun.   The quantities to be meafured are fo fmall, and
the obfervations of the center of Mars fo precarious, and
attended with fuch difficulties (fince an error of only a few
feconds would be fatal) that we muft have recourfe to other
methods.

When I found that the poles of Mars were diftinguifhed with
remarkable luminous fpots *, it occurred to me, that we might
obtain a good theory for fettling the inclination and nodes of
that planet's axis, by meafures taken of the fituation of thofe
fpots.   But, not to proceed upon grounds that wanted confir-
mation, it became neceffary to determine by obfervation, how
far thefe polar fpots might be depended upon as permanent ;
and in what latitude of the globe of Mars they were fituated ;
for, if they fhould either be changeable, or not be at the very
poles, we might be led into great miftakes by overlooking thefe
circumftances.   The following obfervations will affift us in
the inveftigation of thefe preliminary points.

---

* A bright fpot near the fouthern pole, appearing like a polar zone, has alfo
been obferved by M. Maraldi.   See Dr. Smith's Optics, § 1094.

1777, April 17. 7 h. 50′. There are two remarkable bright
fpots on Mars. In fig. 1. tab. VI. they are marked
a and *b*. The line AB expreffes the direction of a
parallel of declination. 10 feet reflector, 9 inches
aperture, power 211 *.
10 h. 20′. They are both quite gone out of the difk.
1779, This year, in all my obfervations on Mars, there is no
mention of any bright fpots, fo that I believe there
were none remarkable enough to attract my attention. However, as my view was particularly directed
to the phænomena of this planet's diurnal rotation,
it is poffible I might overlook them.
1781, March 13. 17 h. 40′. 20 feet reflector. I faw a very
lucid fpot on the fouthern limb of Mars of a confiderable extent. See fig. 2.
June 25. 11 h. 36′. 7 feet reflector, power 227. Two
luminous fpots appeared at *a* and *b*, fig. 3.; *a* is
larger than *b*.
12 h. 15′. With 460. *a* is thicker than *b*, but *b* is
rather longer.
13 h. 12′. *a* is grown thicker, and *b* become thinner.
June 27. 11 h. 20′. The two lucid fpots are on Mars.
June 28. 11 h. 15′. They are both vifible; *a*, fig. 4. is
much thicker than *b*.
12 h. 55′. A line joining *a* and *b* does not go through
the center.
June 30. 10 h. 48′. The fpot *a* is vifible. fig. 5.
11 h. 35′. Both fpots are to be feen.

* Phil. Tranf. vol. LXXI. p. 127. and fig. 17.
1781,

1781, July 3. 10 h. 54'. *a* ſeems to be larger than I have
ſeen it, fig. 6.

11 h. 24'. *b* is not yet viſible, fig. 7.

12 h. 36'. I perceive part of *b*, fig. 8.

July 4. 12 h. 9'. *a* is very full; *b* extremely thin, and
barely viſible.

12 h. 18'. *a* and *b* are not quite oppoſite each other.

12 h. 49'. *b* is increaſed.

July 15. 9 h. 54'. *a* is viſible, fig. 9.

11 h. 35'. *b* invifible.

12 h. 12'. *b* not to be ſeen.

July 16. 11 h. 9'. The bright ſpot *a* is very large.

July 17. 11 h. 15'. No other bright ſpot but *a*.

July 19. 13 h. 31'. *a* viſible.

July 20. 10 h. 3'. I ſuppoſe the bright ſpot *a* on Mars
is, very nearly, the ſouth pole; which therefore
muſt lie in ſight. There is no ſecond bright ſpot *b*
viſible to night.

10 h. 56' *b* not viſible; the night very fine.

July 22. 11 h. 14'. At *a* and *b*, fig. 10. are bright
ſpots; *a* is larger than *b*. Moſt probably the ſouth
pole is in view, and the north pole juſt hid from our
ſight. If the ſpots are polar, or nearly ſo, then *a*
muſt, on a ſuppoſition of the ſouth pole's being in
view, appear larger than *b*; and if *b* extend a little
more from the north pole one way than another, it
muſt be ſubject to ſome change in its appearance from
the revolution of Mars on its axis.

July 30. 9 h. 43'. Both ſpots viſible.

Auguſt 8. 10 h. 4'. Only *a* viſible, fig. 11.

Auguſt 17. 9 h. 21'. Only *a* in ſight.

1781, Auguſt 23. 8 h. 44′. *a* as uſual, and part of *b* viſible,
      fig. 12.

Sept. 7. The white ſpot *a* is very large.

1783, May 20. Mars has a ſingular appearance.   At *a*, fig. 13.
      is the polar ſpot, which is bright, and ſeems to pro-
      ject above the diſk by its ſplendour, cauſing a break
      at *c*.

July 4. *a* is very bright.

July 23. 14 h. 45′. *a* is very lucid.

Auguſt 16. I ſaw the bright ſpot with the 20 feet
      reflector as uſual.

Aug. 26. The lucid ſpot on Mars is its ſouth pole, for
      it remains in the ſame place, while the dark equato-
      rial ſpots perform their conſtant gyrations: it is
      nearly circular.

Aug. 29. The ſouth polar ſpot is in the ſame ſituation.

Sept. 9. As uſual.

Sept. 22. The ſouth polar ſpot is of a circular ſhape,
      and very brilliant and white.   I had a beautiful and
      diſtinct view of it when it was about the meridian,
      and meaſured its little diameter in the equatorial di-
      rection of Mars.   With a power of 932 it gave
      1′ 41‴, and I ſaw it very diſtinctly.   The outward
      edge of the ſpot came juſt up to the upper limb; a
      favourable hazineſs, taking off every troubleſome
      ray, gave me objects in general exceedingly well de-
      fined, eſpecially Mars.

Sept. 23. 9 h. 55 . The polar ſpot *a*, fig. 14. as uſual.

Sept. 24. The ſame.

1783.

1783, Sept. 25. 12 h. 30'. The bright fouth polar fpot *a*, fig.
15. feems to be fixed in its place, and goes nearly up
to the margin of the difk; it is perfectly round.

12 h. 55'. The track of the equatorial fpots is incur-
vated, being convex towards the north, fee *e*, *q*, fig.
23.: this confirms the white fpot's being at the fouth
pole. With long attention I can perceive the edge
of the difk of Mars beyond the fpot, extending about
¼ diameter of the fpot.

Sept. 26. 12 h. 10'. The fpot *a* is in a line with the
center and the end of the hook, fig. 16.

Sept. 27, 28, 29. The fpot as ufual.

Sept. 30. 10 h. 30'. The polar fpot as in fig. 17.

Oct. 1. 9 h. 55'. I am inclined to think, that the white
fpot has fome little revolution, and therefore is not
with its center exactly at the pole of Mars; it is
rather probable, that the real pole, though within
the fpot, may lie near the circumference of it, or
one-third of its diameter from one of the fides. A
few days more will fhew it, as I fhall now fix my
particular attention upon it.

10 h. 17'. The bright fpot is certainly not fo far upon
the difk as it ufed to be formerly, and is either
reduced or has a fmall motion; which of the two
is the cafe will be feen in a few hours.

13 h. 3'. The bright fpot has a little motion; for it is
now come farther into the difk.

I concluded now, in general, that none of the bright fpots
on Mars were exactly at the poles, though they could certainly
not be far from them: for what has been juft related of the

rft,

rft, 2d, and 3d of October 1783, fhews plainly, that the appearance of the fouthern fpot *a* was a little affected by the diurnal motion of the planet; and the obfervations of the 3d and 4th of July 1781, fhew alfo that the fpot *b* could not be exactly at the north pole; and that, perhaps, the vifible branch of the latter extended pretty far towards the equator. However, the fouth polar fpot of the year 1783, being very fmall and nearly round, afforded a good opportunity for determining its polar diftance, by noting the different angles of pofition it affumed while Mars revolved on its axis; to this end many obfervations were taken at different hours of the fame night, which will be found among the meafures of the angles of pofition in the next divifion of my fubject. And fince the different degrees of brilliancy, as well as the proportional apparent magnitude of the fpot, would alfo contribute to the inveftigation of this point, I continued my remarks on thofe particulars, as follows.

1783, Oct. 2. 7 h. 59'. The bright fpot near the fouth pole is about half vifible.

Oct. 4. 8 h. 0'. The polar fpot feems to project above the difk as formerly, and is very fmall.

Oct. 5. 11 h. 13'. The fpot is very fmall, and feems actually to be in the circumference.

11 h. 30'. The fpot is fmall, and feems to be with its fartheft fide in the circumference of the difk; or it may, perhaps, be partly beyond it, and therefore not all in fight.

11 h. 50'. I fee the fpot much clearer than I did before.

13 h. 15'. The white fpot is more in fight, and of its ufual fize, but does not feem much to change its pofition;

ſition; however, what change there is ſhews that it has been beyond the pole, as it appears to have been direct while the equatorial ſpots were retrograde.

1783, Oct. 9. 11 h. 48′. The white polar ſpot increaſes in ſize. At 10 h. 35′. it was as in fig. 18. but is now larger, and coming round towards that part of its orbit which is neareſt to us. See fig. 24.

Oct. 10. 6 h. 20′. I ſee no white polar ſpot; but the planet is too low for any obſervation to be depended on.

6 h. 55′. The white ſpot begins to be viſible; at leaſt I ſee it now, the planet being higher than before, fig. 19.

9 h. 55′. With 460, the white ſpot is conſiderably in-creaſed, and ſhews a circular form, fig. 20.

Oct. 11. 7 h. 46′. The bright ſpot is very viſible; the evening fine; with 278.

Oct. 16. 7 h. 7′. The ſpot is very luminous.

9 h. 55′. It ſeems rather lengthened; perhaps it may be arrived at the extreme of its parallel of decli-nation.

Oct. 17. 7 h. 47′. The white ſpot *a*, fig. 21. is very bright.

13 h. 7′. It is leſs in appearance than it was in the be-ginning of the evening.

Oct. 23. 6 h. 46′. The bright ſpot is very large and luminous; I ſuppoſe it to be in the nearer parts of its little orbit.

7 h. 11′. It is ſituated as in fig 22.

Oct. 24. 7 h. 1′. The white ſpot is very large.

Oct. 27. 8 h. 45′. It is very large and round.

Nov. 1. 7 h. 47′. The ſpot is round and bright.

1783, Nov. 11. The deficiency of light which occasions Mars
to appear gibbous, reaches over the fouth polar fpot
towards the preceding limb, and hides it.

Nov. 14. Mars is gibbous, and the polar fpot is thereby
rendered invifible.

Nov. 17. 6 h. 0'. The fouth polar fpot is under the fal-
cated defect of light.

6 h. 30'. I do not know whether there be not a faint
glimpfe of the polar fpot left; the weather is too
bad to determine it.

I have added fig. 25. (tab. X.) to fhew the connection of the
15th, 17th, 18th, 19th, 20th, 21ft, and 22d figures, which
complete the whole equatorial circle of appearances on Mars,
as they were obferved in immediate fucceffion. The center of
the circle marked 17 is placed on the circumference of the
inner circle, by making its diftance from the center of the
circle, marked 15, anfwer to the interval of time between the
two obfervations, properly calculated and reduced to fidereal
meafure. The fame has been done with regard to the circles
marked 18, 19, &c. And it will be found, by placing any
one of thefe connected circles, fo as to have its contents in a
fimilar fituation with the figures in the fingle reprefentation
which bears the fame number, that there is a fufficient refem-
blance between them; but fome allowance muft undoubtedly
be made for the unavoidable diftortions occafioned by this kind
of projection.

In order to bring thefe obfervations on the bright fpots into
one view, I have placed them at the circumference of three
circles (fee fig. 26, 27, 28. tab. VII. VIII. IX.) divided into degrees, reprefenting the parallels of declination in which they

revolved about the poles of Mars. The divifion of the circles marked 360 is where a fpot paffes that meridian of the planet which is turned towards the earth, and where, confequently, it appears to us in its greateft luftre. The motion of the fpot is according to the numbers 30, 60, 90, and fo on to 360. In calculating the daily places of the fpots I have ufed the fidereal period of 24 h. 39′ 21″,67 determined in my paper on the rotation of Mars[*]; and have alfo made proper allowances for the alterations of the geocentric longitudes calculated from the fituations of that planet given in the Nautical Almanack; by which means the fidereal is reduced to a proper fynodical period.

The following three tables contain the refult of the calculations, and ferve to explain the arrangement of the obfervations in the circles. In the firft column are the times when the obfervations were made. In the fecond, the fidereal places of the fpot in degrees and minutes. In the third column are the geocentric longitudes of Mars at the time of the obfervations. In the fourth, the neceffary corrections on account of thefe different longitudes. In the fifth column are the corrected or fynodical places of the fpots; and, according to the numbers in that column, they are marked on the circles, where confequently each fpot is reprefented as it muft have appeared to be fituated at the time of obfervation.

[*] Phil Tranf. vol. LXXI. p 134.

TABLE

T A B L E I.

Time of observation.			Sider. place.		Geoc. longit.			Correction.		Synod. place.	
D.	H.	M.	D.	M.	S.	D.	M.	D,	M.	D.	M.
June 25	11	36	359	31	9	24	35	+0	0	350	31
25	12	15	0	0	9	24	35	+0	0	0	0
25	13	12	13	52	9	24	34	−0	1	13	51
27	11	20	357	28	9	24	12	−0	23	357	5
28	11	15	316	40	9	24	1	−0	34	316	6
30	10	48	290	56	9	23	38	−0	57	289	59
30	11	35	302	23	9	23	38	−0	57	301	26
July 3	10	54	263	40	9	22	55	−1	40	262	0
3	11	24	270	58	9	22	55	−1	40	269	18
3	12	10	282	9	9	22	55	−1	40	280	29
3	12	36	288	29	9	22	55	−1	41	286	48
4	12	9	272	20	9	22	40	−1	55	270	25
4	12	49	282	4	9	22	40	−1	55	280	9
15	9	54	134	7	9	19	43	−4	52	129	15
15	11	35	158	42	9	19	42	−4	53	153	49
15	12	12	167	42	9	19	42	−4	53	162	49
16	11	9	142	48	9	19	26	−5	9	137	39
17	11	15	134	40	9	19	9	−5	26	129	14
19	13	31	148	37	9	18	34	−6	1	142	36
20	10	3	88	25	9	18	21	−6	14	82	11
20	10	56	101	19	9	18	20	−6	15	95	4
22	11	14	86	32	9	17	49	−6	46	79	46
30	9	43	347	46	9	16	5	−8	30	339	16

TABLE II.

Time of observation.			Sider. place.		Geoc. longit.			Correction.		Synod. place.	
	D.	H.	M.	D.	M.	S.	D.	M.	D.	M.	D.	M.
June 25	11	36	86	51	9	24	35	+ 1	40	88	31
25	12	15	96	20	9	24	35	+ 1	40	98	0
25	13	12	110	12	9	24	34	+ 1	39	111	51
28	11	15	53	0	9	24	14	+ 1	6	54	6
30	10	48	27	16	9	23	38	+ 0	43	27	59
30	11	35	38	43	9	23	38	+ 0	43	39	26
July 3	10	54	0	0	9	22	55	+ 0	0	0	0
4	12	9	8	40	9	22	40	– 0	15	8	25
15	9	54	230	27	9	19	43	– 3	12	227	15
15	10	12	234	50	9	19	43	– 3	12	231	38
15	11	35	255	2	9	19	42	– 3	13	251	49
15	12	12	264	2	9	19	42	– 3	13	260	49
16	11	9	339	8	9	19	26	– 3	29	235	39
19	13	31	244	57	9	18	34	– 4	21	240	36
20	10	3	184	45	9	18	21	– 4	34	180	11
20	10	56	197	39	9	18	20	– 4	35	193	4
30	9	43	84	6	9	16	5	– 6	50	77	16

TABLE

T A B L E . III.

Time of obfervation.			Sider. place.		Geoc. longit.			Correction.		Synod. place.	
D.	H.	M.	D.	M.	S.	D.	M.	D.	M.	D.	M.
Sept. 25	13	30	6	32	0	9	54	+6	44	13	16
Oct. 1	10	17	262	5	0	8	6	+4	56	267	1
1	13	3	302	29	0	8	5	+4	55	307	24
2	7	59	218	55	0	7	50	+4	40	223	25
4	8	0	200	0	0	7	15	+4	5	204	5
4	8	46	211	12	0	7	15	+4	5	215	17
5	11	13	237	23	0	6	55	+3	45	241	8
5	11	30	241	31	0	6	55	+3	45	245	16
5	11	50	246	23	0	6	55	+3	45	250	8
5	13	15	267	4	0	6	54	+3	44	270	48
5	14	0	278	1	0	6	53	+3	43	281	44
7	8	20	176	8	0	6	23	+3	13	179	21
7	10	5	201	41	0	6	22	+3	12	204	53
7	11	50	227	14	0	6	21	+3	11	230	25
9	11	48	207	35	0	5	49	+2	39	210	14
10	6	55	126	42	0	5	37	+2	27	129	9
10	7	50	140	5	0	5	36	+2	26	142	31
10	9	55	170	30	0	5	34	+2	24	172	54
10	12	11	203	36	0	5	33	+2	21	205	57
16	7	7	72	9	0	4	15	+1	5	73	14
16	7	46	81	39	0	4	15	+1	4	82	43
16	9	55	113	2	0	4	14	+1	4	114	6
17	7	47	72	19	0	4	3	+0	53	73	12
17	13	7	150	11	0	4	0	+0	50	151	1
23	6	46	0	0	0	3	10	−0	0	0	0
24	7	1	354	0	0	3	2	−0	8	353	52

From the appearance and difappearance of the bright north polar fpot in the year 1781, we collect that the circle of its motion, reprefented by fig. 26. was at fome confiderable diftance from the pole. By a calculation, made according to the principles hereafter explained, its latitude muft have been about 76° or 77° north; for I find that, to the inhabitants of Mars, the declination of the fun, June 25. 12 h. 15′ of our time, was about 9° 56′ fouth *; and the fpot muft have been at leaft fo

* See p. 259. and 260.

K k 3

far

far removed from the north pole as to fall a few degrees within the enlightened part of the diſk, to become viſible to us.

The ſouth pole of Mars could not be many degrees from the center of the large bright ſouthern ſpot of the year 1781, whoſe courſe is traced in fig. 27 ; though the ſpot was of ſuch a magnitude as to cover all the polar regions farther than the 70th or 65th degree, and in that part which was on the meridian July 3, at 10 h. 54′, perhaps a little farther.

In the next diviſion of our ſubject will be ſhewn, that the inclination and poſition of the axis of Mars are ſuch, that the whole circle, fig. 28. (which will appear to be in about 81° 52′ of ſouth latitude on the globe of Mars) was in view all the time the obſervations on the bright ſouth polar ſpot of the year 1783, which are marked upon it, were made, but in ſo oblique a ſituation as to be projected into a very narrow ellipſis. See fig. 24. where *m n* is the little ellipſis in which the ſpot *a* revolved about the pole. Hence then we may eaſily account for the obſerved magnitude and brightneſs of the ſpot Oct. 23, 24, and 27. when it was expoſed to us in its meridian ſplendour. Its ſituations Oct. 16. and 17. on one extreme of the parallel, as well as thoſe of Oct. 5. and Nov. 1. on the other, gave us alſo a bright view of it : and, when we paſs over to that half of the circle which lies beyond the pole, the much greater obliquity into which the ſpot muſt there be projected will perfectly account for its being ſmaller at 13 h. 7′ of Oct. 17. than at 7 h. 47′ of the ſame evening. It will alſo explain its ſmallneſs Oct. 4. and its increaſe Oct. 9. We ſhall have occaſion hereafter to recur to the ſame figure, ſo that I take no notice at preſent of the angles of poſition which are marked upon it.

4   *Of*

Of the direction or nodes of the axis of Mars, *its inclination to the ecliptic, and the angle of that planet's equator with its own orbit.*

From the foregoing article we may gather, that the bright polar fpots on Mars are the moft convenient objects for determining the fituation of the axis of this planet; I fhall therefore collect, in one view, all the meafures I have taken of thefe fpots for that purpofe. Before I conftructed a micrometer for taking the angle of pofition, I ufed to draw a line through the figure delineated of Mars to reprefent the parallel of declination; in a few of my firft obfervations, therefore, I can only take the fituation of the polar fpots from fuch drawings, and of confequence no great accuracy in the angles, as to the exact number of degrees, can be expected.

1777, April 17. 7 h. 50'. A line drawn through the middle of the two bright polar fpots *a* and *b*, fig. 1. makes an angle of about 63°, with a parallel of declination AB; the fouthern fpot preceding and the northern following.

My reafon for chufing a line drawn through both the fpots rather than through one of them and the center is, firft, that they were not fituated quite oppofite each other, and therefore, unlefs other obfervations had pointed out which was moft polar, I fhould evidently run the greater rifk in fixing on one of them in preference to the other. In the next place, we find by the fecond obfervation, page 235. that in two hours and a half both fpots were intirely gone out of the difk. This plainly

plainly denotes, that they were both in the same half of a sphere orthographically projected, and divided by a plane passing through the axis of Mars and the eye, but that neither of them were polar. Now, a line drawn through two points not far from opposite each other, both in the same hemisphere, and both removed from the poles of it, must approach more to a parallelism with the axis, than a line drawn through either of them and the center.

1779, May 9. There being no bright spots by which to judge of the position of the poles, it is estimated from a well known dark equatorial spot, with a line drawn through the figure to denote a parallel of declination. By very rough estimation it is about 42° south preceding.

May 11. The same figure, being drawn again in another situation, and also with a line giving a parallel of declination, points out, by the same rough estimation, 62° south preceding.

1781, June 25. 11 h. 35'. The position of the spots *a* and *b*, fig. 3. with regard to a parallel of declination, measured with a micrometer 74° 32'. The spot *a* was south preceding, and *b* north following.

July 15. 10 h. 12'. The angle of position, of the center of the spot *a*, fig. 9. through the center of the disk, 74° 18' south preceding.

1783, August 16. Position of the spot *a*, 64° south following the center; but as the planet is not full, the center becomes dubious, and the measure therefore may not be quite accurate, though taken with a 20 feet reflector; power 200.

Sept.

1783, Sept. 9. Pofition of the fuppofed fouth pole of Mars 65° 12' fouth following; 7 feet reflector; power 460.

Sept. 22. Pofition of the fame 52° 9' f. following; 460.

Sept. 25. 13 h. 30'. Pofition of the fouth polar fpot 56° 27'. very accurately taken, by bifecting the difk of Mars through the bright fpot, and fuppofing the planet now near enough the oppofition to induce no material error. Hitherto I have taken it through a fuppofed center by endeavouring to allow a little for what I thought the deficiency in the difk; but not to-night.

Oct. 4. 8 h. 46'. Pofition of the fpot 51° 21'; Mars too low and hazy to depend much on the meafure with fo high a power as 460.

Oct 5. The motion of the polar fpot being now ftrongly fufpected, or rather already known, I took the following meafures, by way of difcovering its quantity.

11 h. 50'. Pofition very exactly taken 50° 6' f. following.

14 h. 0'. Pofition of the fpot 49° 45'.

Oct. 7. 8 h. 20'. Pofition 55° 12'. In order to fee how far this meafure might be trufted to, I fet 49° 36' in the micrometer, which was evidently too fmall; next I took 51° 36', which was alfo too fmall; after this, I took a new meafure, and found 55° 24', which appeared to me very exact. 10 h. 5'. The pofition now was 53°. 11 h. 50'. It meafured 52° 12'. As there is nothing to diftinguifh the center, it is extremely difficult to pleafe one's felf in bringing the fpot into a line with it.

1783, Oct. 10. 7 h. 50′. Position of the polar spot 57° 12′; with 460, very accurate. I tried a few parts less of the micrometer, but found the measure too little. I see pretty distinctly, but the air is tremulous.

9 h. 55′. Position 52° 42′; very distinct.

12 h. 11′. Position 46° 30′; I see not quite so well now as I could wish.

14 h. 1′. Position 44° 12′; but liable to great uncertainty, on account of tremulous air; it becomes more difficult to distinguish the center when the planet is not perfectly defined.

Oct. 16. 7 h. 7′. Position 63° 9′. By way of trial I set 59° 36′; which was too small; also 60° 24′ was too small; again, 61° 24′ was not large enough. Then, taking a fresh measure, I found it 62° 48′, which I thought right.

9 h. 55′. I took three measures, and thought the third, which was 65° 0′, the best of all; for I saw the planet and the spot remarkably well.

Oct. 27. 8 h. 45′. Position of the polar spot 59° 30′. I took three other measures, of which 60° 39′ appeared to me the best; it was taken with long attendance and many changes and trials of the wires in different positions; but the gibbosity of Mars is such, that measures of the situation of the spot are now no longer to be depended on.

These positions, I believe, will be sufficient for the purpose of settling the latitude of the polar spots, and thereby obtaining a correct measure of the situation of the real pole. I have referred those of the south polar spot of the year 1783 to the same circle which contains the observations that were made on

2   the

the apparent brightnefs and magnitude of that fpot, that they may be compared together. (See fig. 28.) The agreement of the meafures, and the phænomena attending the motion of the fpot, are fufficient to point out the meridian of the circle; for which, from a due confideration of thefe circumftances, I have fixed on the place where the fpot was Oct. 10. 6 h. 46'.

Of the angles collected in fig. 28. we find 65° 0' the largeft, and 49° 45' the fmalleft; but, on account of the different fituation of the earth and Mars, the angle meafured 7' lefs Oct. 16. than it would have done had the planets remained in the places they were in Oct. 5. when the other meafure was taken. This being added, we have 65° 7'. The difference between the two pofitions is 15° 22'. Now, the conftruction of fig. 28. being admitted, we fee that the angles were nearly taken at the oppofite extremes of the circle in which the fpot moved. However, by the 5th column of Tab. III. Oct. 5. we have the fituation of the fpot in the circle with refpect to the meridian 281° 44', and Oct. 16. 114° 6': therefore the fouth polar diftance of the center of the fpot is found, by taking half the fum of the fines of thefe angles to radius, as 7° 41' (half of 15° 22') to a fourth number, which is 8° 8'; and the latitude of the circle, in which the fpot moved about the pole, therefore is 81° 52' fouth. This being determined, we have the following correction for the angles of pofition: radius is to fine of the angular diftance of the fpot from the meridian as 8° 8' to the required quantity. This muft be added or fubtracted, according as the cafe requires; and thereby we fhall have the pofition of the true pole from any one of the meafures.

I fhall now apply the above to determine the fituation of the axis of Mars. To this end, we fee that, in the firft place, the

meafures

meaſures muſt be correćted for the latitude of the ſpot; next, they muſt be reduced to a heliocentric obſervation, which will alſo correćt them from the difference occaſioned by the different ſituation of the planets when they were taken. This being done, we may ſelećt two obſervations at a proper diſtance; from which, by trigonometry, we ſhall have the node and inclination of the axis. When theſe elements are obtained, it will be eaſy to ſee how other obſervations agree with them; which will afford the means of correćting or verifying the former calculations.

Let T, fig. 29. (tab. X) be the earth; ♋ Q *q* ♑ the ecliptic as ſeen from T; P the point of the heavens towards which the north pole of the earth is direćted; M the place of the orbit of Mars *μ m* M, where an obſervation of the poles of that planet has been made, which is to be reduced to its heliocentric meaſure. And, firſt, ſuppoſe it to have been made at the time of the oppoſition of that planet. Then, the place M or Q in the ecliptic being given, we have the ſides Q ♋, ♋ P; whence the angle Q, of the right-angled triangle P ♋ Q, is found. This being added to, or taken from, the obſerved angle of poſition of the axis of Mars, according to circumſtances eaſily to be determined, reduces it to its heliocentric poſition. But if this obſervation was not made at the time of an oppoſition, but at ſome other place *m*, a ſecond correćtion is to be applied in the following manner.

Let the angle *q*, of the triangle P ♋ *q*, be found as before, and properly applied to the poſition of the axis of Mars now at *m*; then make the angle *m* S *μ*, at the ſun S, equal to the angle S *m* T, and *μ* will be the heliocentric place, where the angle of poſition, when ſeen from S, will appear to be as it was found at *m*, after the application of the firſt correćtion:

4   for

for S μ being parallel to T m, and fuppofing the axis of Mars
to preferve its parallelifm while it moves from m to μ, appearances of Mars at μ to an eye at S, muft be the fame as they
are at m to an eye at T.

The following table contains the refult of calculations relating to the angles of fig. 28. In the firft column are the
times when the obfervations were made. In the fecond, the
angles as they were taken. In the third column are the quantities of the angles Q, q, calculated from the geocentric longitudes contained in the third column of the third table. In
the fourth column are the corrections for the fituation of the
fpot in the circle of latitude obtained from the fines of the
angles in the fifth column of the third table. In the fifth
are the corrections requifite on account of the change of fituation of the planets, during the interval between the feveral
days on which the meafures were taken; thefe are obtained
from the third column of this table, and I have affumed the
4th of October, as being the obfervation neareft the oppofition, to which I have reduced the other meafures. In the fixth
column are the angles of the fecond, corrected by the quantities contained in the fourth and fifth columns, applied according to their figns.

TABLE

T A B L E IV.

Time of observation.			Angles taken.		Angle Q.		First correction.		Second correct.	Angles corrected.	
D.	H.	M.	D.	M.	D.	M.	D.	M.	M.	D.	M.
Sept.25	13	30	56	27	+23	10	—1	52	—8	54	27
Oct. 4	8	46	51	21	+23	18	+4	42	—0	56	3
5	11	50	50	6	+23	19	+7	39	+1	57	46
5	14	0	49	45	+23	19	+7	59	+1	57	45
7	8	20	{55 12 / 55 24}		+23	21	—0	7	+2	{55 7 / 55 19}	
7	10	5	53	0	+23	21	+3	26	+3	56	29
7	11	50	52	12	+23	21	+6	16	+3	58	31
10	7	50	57	12	+23	22	—4	57	+4	52	19
10	9	55	52	42	+23	22	—1	7	+4	51	39
16	7	7	{63 9 / 62 48}		+23	25	—7	47	+7	{55 29 / 55 8}	
16	9	55	65	0	+23	25	—7	23	+7	57	45

As we have no particular reason to select one measure rather than another, a mean of all the 13 will probably be nearest the truth; so that by these observations, which, as we said before, are reduced to the 4th of October, 1783, we find the position of the axis of Mars that day to have been 55° 41' south following.

From the appearances of the south polar spot in 1781, represented fig. 27. we may conclude, that its center was nearly polar. We find it continued visible all the time Mars revolved on its axis; and, to present us generally with a pretty equal share of the luminous appearance, a spot which covered from 45° to 60° of a great circle on the globe of Mars could not have any considerable polar distance: however, a small correction in the angle of position seems to be necessary, which should be taken from the measure of the 15th of July, because that branch of the spot which probably extended farthest towards

the

the equator, was then in the *following* quadrant. The meafure of both the fpots on June the 25th, 1781, is ftill more
to be depended on, as giving us very nearly the pofition of the
true pole; for it appears evident from the phænomena of the
bright north-polar fpot in fig. 26. that that fpot was in the
meridian when the meafure was taken, while the fouthern fpot
was in the *preceding* quadrant near its greateft limit. Now,
fince an angle at the circumference of a circle is but half the
angle at the center, when the arches which fubtend thefe
angles are equal, the correction neceffary to be applied to the
meafure taken through the two fpots will be but one half of
the correction which would have been requifite had it been
taken through the center; therefore, in order to reduce this
to the condition of the former, we may fuppofe it to have been
taken through the center of Mars when the fpot was only 30, or
150 degrees from the meridian. It is alfo neceffary to add 1° 54′ to
the angle of July 15, which it would have meafured more had
the planets remained where they were June 25. This done,
we may have the polar diftance of the center of the fpot as
before. Half the fum of the fines (of 231° 38′ and 150°) to
radius, as 50′ (half the difference between 74° 32′ and 76° 12′)
to a fourth number, which is 1° 18′.

I fhould obferve here, that the meafures of the angle of pofition would be too large before the fpot came to the meridian,
and too fmall afterwards, the axis of Mars being fouth preceding; whereas, in fig. 28. they would be too fmall before,
and too large after, the meridian paffage, the pole being fouth
following.

Thefe two obfervations arranged as thofe in the fourth table,
and reduced to the time of the 25th of June, will ftand as
follows.

TABLE

T A B L E V.

Time of obfervation.			Angles taken.		Angle Q.		First correction.		Second correct.		Corrected Angle.	
D.	H.	M.	D.	M.	D.	M.			D.	M	D.	M.
June 25	11	36	74	32	−10	14	+	{ half of	−0	0	75	11
								{ 1° 18′				
July 15	10	12	74	18	− 8	20	−1	1	+1	54	75	11

I am to remark, that we have here admitted both meafures as equally good; and that, therefore, the refult is a mean of them both, and fhews the axis of Mars, June 25, 1781, to have been 75° 11′ fouth preceding.

Our next bufinefs will be to reduce thefe two geocentric ob-fervations to a heliocentric meafure. This is to be done, as we have fhewn before, by a calculation of the angle Q, fig. 29. The refult of it fhews, that 10° 14′ are to be fubtracted from the mean corrected angle of pofition, reduced to June 25, 1781, and 23° 18′ to be added to the angle which is the corrected mean of 13 meafures, reduced to Oct. 4, 1783. Hence we learn, that on thofe days and hours, when the heliocentric places of Mars were 9 s. 24° 35′, and 0 s. 7° 15′ (which would happen about July 18, 1781, and Sept. 29, 1783) an ob-ferver placed in the fun would have feen, on the former, the axis of Mars inclined to the ecliptic 64° 57′, the north pole being towards the left; and on the latter, he would have feen the fame axis inclined to the ecliptic 78° 59′, the north pole being then towards the right.

The firft conclufion we may draw from thefe principles is, that the north pole of Mars muft be directed towards fome point of the heavens between 9 s. 24° 35′ and 0 s. 7° 15′; be-caufe the change of the fituation of the pole from left to right, which

which happened in the time the planet paffed from one place to the other, is a plain indication of its having gone through the node of the axis. Next, we may alfo conclude, that the node muft be confiderably nearer the latter point of the ecliptic than the former; for, whatever be the inclination of the axis, it will be feen under equal angles at equal diftances from the node.

But, by a trigonometrical procefs of folvinga few triangles, we foon difcover both the inclination of the axis, and the place where it interfects the ecliptic at rectangles (which, for want of a better term, I have perhaps improperly called its node). Accordingly I find, by calculation, that the node is in 17° 47′ of Pifces, the north pole of Mars being directed towards that part of the heavens; and that the inclination of the axis to the ecliptic is 59° 42′.

We fhall now compare the obfervations of an earlier date with thefe principles, to fee how far they agree. Some of the particulars and calculations relating to them are as follow.

T A B L E VI.

Times of Obfervation.			Eftimations.	Geoc. longit.			Angle Q.		2d correct.
D.	H.	M.	D.	S.	D.	M.	D.	M.	
1779, May 9	12	0	42	7	22	20	+14	45	+ 0
May 11	12	0	62	7	21	40	+15	11	+ 26
1777, Apr. 17	7	50	63	6	3	34	+23	26	

May the 9th, 1779, as we have feen, the angle of pofition was roughly eftimated at 42°, and May 11. at 62°. The great difagreement of thefe coarfe eftimations is undoubtedly owing to the very different fituation of the dark fpot from which they

were taken ; however, ſince we do not mean to uſe theſe obſervations in our calculations, they may ſuffice in a general
way to ſhew, that the axis of Mars was actually about that
time in ſuch a ſituation as our principles give it : for, reducing
the two poſitions to the 9th of May, that of the 11th, from
an allowance of 26′ for the ſituation of the planets, will become 62° 26′ ; and a mean of the two, 50° 13′ ſouth preceding ; which, reduced to a heliocentric obſervation, gives
66° 30′, the north pole lying towards the left. Now, on calculating from the poſition of the node and inclination of the
axis before determined, we find, that the heliocentric angle
was 62° 49′, the north pole pointing towards the left ; and a
nearer agreement with theſe principles could hardly be expected
from eſtimations ſo coarſe. If we go to the year 1777, and
take the poſition of the two bright ſpots obſerved the 17th of
April, we have 63° ſouth preceding ; this, reduced to a heliocentric quantity, gives 86° 26′ of inclination, the north pole
being to the left. By calculating we find, that that pole was
then actually 81° 27′ inclined to the ecliptic, and pointed
towards the left as ſeen from the ſun.

The inclination and ſituation of the node of the axis of Mars,
with reſpect to the ecliptic being found may thus be reduced
to that planet's own orbit. Let EC, fig. 30. (tab. X.) be a part
of the ecliptic ; OM part of the orbit of Mars ; PEO a line
drawn from P, the celeſtial pole of Mars, through E, that
point which has been determined to be the place of the node of
the axis of Mars in the ecliptic, and continued to O where it in-
terſects the orbit of Mars. Now, if according to Mr. DE LA
LANDE we put the node of the orbit of Mars for 1783, in
1 s. 17° 58′, we have from the place of the node of the axis
(that is, 11 s. 17° 47′) to the place of the node of the orbit,

an arch EN of 60° 11′; in the triangle NEO, right-angled at E, there is also given the angle ENO, according to the same author, 1° 51′, which is the inclination of the orbit of Mars to the ecliptic. Hence we find the angle EON 89° 5′, and side ON 60° 12′. Again, when Mars is in the node of its orbit N, we have, by calculation from our principles, the angle PNE = 63° 7′, to which, adding the angle ENO = 1° 51′, we have PNO = 64° 58′; from which two angles PON and PNO with the distance ON, we obtain the inclination of the axis of Mars, and place of its node with respect to that planet's own orbit; the inclination being 61° 18′, and the place of the node of the axis 58° 31′ preceding the intersection of the ecliptic with the orbit of Mars, or in our 19° 28′ of Pisces.

Being thus acquainted with what the inhabitants of Mars will call the obliquity of their ecliptic, and the situation of their equinoctial and solstitial points, we are furnished with the means of calculating the seasons on Mars; and may account, in a manner which I think highly probable, for the remarkable appearances about its polar regions.

But first it may not be improper to give an instance how to resolve any query concerning the martial seasons. Thus, let it be required to compute the declination of the Sun on Mars, June 25, 1781, at midnight of our time. If ♈ ♉ ♊ ♋, &c. fig. 31. (tab. X.) represent the ecliptic of Mars, and ♈ ♋ ♎ ♑ the ecliptic of our planet, A*a*, *b*B, the mutual intersection of the martial and terrestrial ecliptics, then there is given the heliocentric longitude of Mars, ♈*m* = 9 s. 10° 30′; then taking away six signs, and ♎ *b*, or ♈*a* = 1 s. 17° 58′, there remains *bm* = 1 s. 22° 32′. From this arch, with the given inclination, 1° 51′, of the orbits to each other, we have cosine of inclination to radius, as tangent of *bm* to tangent of BM = 1 s. 22° 33′. And

M m 2   taking

taking away B ♈ = 1 s. 1° 29′, which is the complement to ♑ B (or ♋ A, already ſhewn to be 1 s. 28° 31′) there will remain ♈ M = 0 s. 21° 4′, the place of Mars in its own orbit *; that is, on the time abovementioned, the ſun's longitude on Mars will be 6 s. 21° 4′, and the obliquity of the martial ecliptic 28° 42′ being alſo given, we find, by the uſual method, the ſun's declination 9° 56′ ſouth.

The analogy between Mars and the earth is, perhaps, by far the greateſt in the whole ſolar ſyſtem. Their diurnal motion is nearly the ſame; the obliquity of their reſpective ecliptics, on which the ſeaſons depend, not very different; of all the ſuperior planets the diſtance of Mars from the ſun is by far the neareſt alike to that of the earth: nor will the length of the martial year appear very different from that which we enjoy, when compared to the ſurpriſing duration of the years of Jupiter, Saturn, and the Georgium Sidus. If, then, we find that the globe we inhabit has its polar regions frozen and covered with mountains of ice and ſnow, that only partly melt when alternately expoſed to the ſun, I may well be permitted to ſurmiſe that the ſame cauſes may probably have the ſame effect on the globe of Mars; that the bright polar ſpots are owing to the vivid reflection of light from frozen regions; and that the reduction of thoſe ſpots is to be aſcribed to their being expoſed to the ſun. In the year 1781, the ſouth polar ſpot was extremely large, which we might well expect, ſince that pole had but lately been involved in a whole twelve-month's darkneſs and abſence of the ſun; but in 1783 I found it conſiderably ſmaller than before, and it decreaſed continually

---

* If no very great accuracy be required, we may add 3 s. 10° 34′ to any given place of our ecliptic, which will at once reduce it to what it ſhould be called on the orbit of Mars, and will always be true to within a minute.

from the 20th of May till about the middle of September, when it feemed to be at a ftand. During this laft period the fouth.pole had already been above eight months enjoying the benefit of fummer, and ftill continued.to receive the fun-beams; though, towards the latter end, in fuch an oblique direction as to be but little benefited by them.. On the other hand, in the year 1781, the north polar fpot, which had then been its twelve-month in the fun-fhine, and was but lately returning to darknefs, appeared small, though undoubtedly increafing in fize. Its not being vifible in the year 1783 is no objection to thefe phænomena, being owing to the pofition of the axis, by which it was removed out of fight; moft probably, in the next oppofition we fhall fee it renewed, and of confiderable extent and brightnefs; as, by the pofition of the axis of Mars, the fun's fouthern declination will then.be no more than 6° 25' on that planet.

Of the spheroidical figure of Mars.

That a planetary globe, fuch as Mars, turning on an axis, fhould be of a fpheroidical form, will eafily find admittance, when two familiar inftances in Jupiter and the.earth, as well as the known laws of gravitation and centrifugal force of rotatory bodies, lead the way to the reception of fuch doctrines. So far from creating difficulties or doubts, it will rather appear fingular, that the fpheroidical form of this planet, which the following obfervations will eftablifh, has not already been noticed by former aftronomers; and yet, reflecting on the general appearances of Mars, we foon find that opportunities for making obfervations on its real form cannot be very frequent: for, when it is near enough to view it to an advantage, we fee it

6   generally

generally gibbous, and its oppoſitions are ſo ſcarce, and of ſo
ſhort a duration, that in more than two years time we have
not above three or four weeks for ſuch obſervations.   Beſides,
aſtronomers being already uſed to ſee this planet generally
diſtorted, the ſpheroidical form might eaſily be overlooked.

Obſervations relating to the polar flattening of Mars.

1783, Sept 25. 9 h. 50′. I can plainly ſee that the equatorial
    diameter of Mars is longer than the polar.   Meaſure
    of the equatorial diameter 21″ 53‴; of the polar
    diameter 21″ 15‴ *full meaſure*, that is, certainly not
    too ſmall.   The wires were ſet as outward tangents
    to the diſk, and the zero, as well as the meaſures,
    were taken by the light of Mars.

    Sept. 28. 14 h. 25′. I ſhewed the difference of the
    polar and equatorial diameters of Mars to Mr. Wil-
    son, Aſſiſtant Profeſſor of Aſtronomy at Glaſgow.
    He ſaw it perfectly well, ſo as to be entirely con-
    vinced it was not owing to any defect or diſtortion
    occaſioned by the eye lens; and, becauſe I wiſhed
    him to be ſatisfied of the reality of the appearance,
    while he was obſerving, I reminded him of ſeveral
    well known precautions; ſuch as cauſing the planet
    to paſs directly through the center of the field of
    view, and judging of its figure at the time when it
    was moſt diſtinct and beſt defined, and ſo forth.

    Sept. 29. I ſhewed the difference of the polar and equa-
    torial diameters of Mars to Dr. Blagden and Mr.
    Aubert.   Dr. Blagden not only ſaw it imme-
            diately,

diately, but thought the flattening almoſt as much as that of Jupiter. Mr. AUBERT alſo ſaw it very plainly, ſo as to entertain no manner of doubt about the appearance.

As we cannot take too many opportunities of confirming our own obſervations by the eyes of other obſervers, I eſteemed it a very fortunate circumſtance to have the honour of a viſit from theſe gentlemen at ſo particular a time, Mars being this day within 37 hours of the oppoſition, and yeſterday when Mr. WILSON ſaw it, within about two days and a half.

1783, Sept. 30. 10 h. 52′. The difference in the diameters of Mars is very evident and conſiderable.

Meaſure of the equatorial diameter 22″ 9‴ with 278.
Second meaſure - - 22″ 31‴ full large.
Polar diameter very exact - 21″ 26‴.

Oct. 1. 10 h. 50′. I took meaſures of the diameters of Mars with my 20-feet reflector. The equatorial meaſured 103 parts of the micrometer; the polar 98. The value of the diviſions in ſeconds and thirds not being well determined, on account of ſome late change in the focal length of the ſeveral 20-feet object metals I uſe, we have only from theſe meaſures the proportion of the diameters as 103 to 98.

13 h. 15′. Every circumſtance being favourable, I took the following meaſures of the diameters of Mars with my 7-feet reflector, and a diſtinct power of 625.

Equatorial diameter 22″ 12‴ narrow meaſure.
22″ 46‴ rather full.
22″ 35‴ exact.

Polar

Polar diameter   21″ 24‴
21″ 33‴ very exact.

I ſaw Mars perfectly well all the time I meaſured, with all its figures upon the diſk appearing diſtinctly ; and, I think, theſe meaſures may be depended upon better than any I have yet taken.

1783, Oct 5. 14 h. 0′. The difference of the diameters is very ſenſible.

Oct. 7. 9 h. 43′. The flattening of the poles is very viſible.

13 h. 40′. I turned my Newtonian 7-feet reflector one quarter round, ſo as to bring the place to look in at to the bottom; and, as well as the uneaſy poſture would allow, I ſaw the flattening of the poles the ſame as when I looked in at the ſide ; power 460.

14 h. 30′. With a 3½ feet achromatic teleſcope and a ſingle eye lens, I ſaw the difference of the polar and equatorial diameters very plainly.

Oct. 9. 8 h. 40′. I turned my reflector 90° round, ſo as now to look in at the upper end, but ſaw not the leaſt difference in appearances ; for, returning it again immediately to its uſual poſition, in both caſes the equatorial diameter appeared a little longer than the other ; power 278, and the evening fine.

I turned the great ſpeculum one quadrant in its cell, but appearances were not in the leaſt altered; the equatorial diameter ſtill was a little longer than the polar one.

I tried a very fine new object ſpeculum, and found alſo the equatorial diameter a little longer than the polar one.

1783,

1783, Oct. 9. 10 h. 47′. The flattening at the poles very visible.

Oct. 10. 9 h. 55′. A little of the polar flattening is visible, so as to admit of no doubt; power 460, very distinct.

11 h. 32′. Mars visibly flattened, but not much; the achromatic shews it also.

11 h. 42′. The disk of Mars is visibly spheroidical.

Oct. 11. 7 h. 37′. Mars is plainly gibbous, therefore measures and estimations of the diameters must for the future be improper.

11 h. 12′. It is rather difficult to say of what shape Mars is now, for it is partly flattened and partly gibbous; but the gibbous side not being quite in the polar direction of Mars, this produces altogether an odd mixture of shapes: however, upon the whole, the polar diameter is still rather the smallest.

11 h. 13′. The *preceding* side of Mars shews the flattening of the poles, while the *following* is terminated by an elliptical arch.

Oct. 12. 11 h. 12′. The flattening upon the whole is visible.

Oct. 17. 13 h. 7′. The effect of gibbosity is scarcely equal to the flattening; or, upon the whole, the planet is still rather broader over the equator than over the poles.

Nov. 1. 7 h. 56′. The semi-disk, which is *full*, is evidently part of an oblate spheroid; but, to an eye not attentively looking for it, and knowing the shape and exact situation of the poles of Mars, this would probably not appear.

1783, Nov. 10. 9 h. 30′. The gibbofity of Mars is now fuch, that the polar diameter is confiderably longer than the equatorial; but the deficiency not being exactly from pole to pole, makes the difk of a crooked, irregular figure, and renders precifion in this eftimation impoffible; otherwife the phafe of Mars would have made a pretty good micrometer upon the equatorial diameter, and it was with fuch a view I had directed my attention to this circumftance: appearances, however, are vifibly in favour of the polar diameter's being the longeft.

We find that the quick alterations in the vifible difk of Mars, during the time it is in the beft fituation for us to obferve it, are fuch, that if we were to ufe many meafures which have been taken of its diameters, we fhould be obliged to have recourfe to a computation of its phafes, in order to make proper allowance for them. Now, fince thefe changes are in a longitudinal direction, and the poles of Mars are not perpendicular to the ecliptic, it would bring on a calculation of fmall quantities, which it is always beft not to run into where it can be avoided. For this reafon, I fhall at once fettle the proportion of the equatorial to the polar diameter of this planet, from the meafures which were taken on the very day of the oppofition. I prefer them alfo on another account, which is, that they were made in a very fine, clear air, and were repeated with a very high power, and with two different inftruments, of whofe faithful reprefentation of celeftial objects, the many obfervations on very clofe double ftars I have made with them have given me very evident proofs.

As

As we are at prefent only in queft of the proportion of one diameter to the other, the meafures of the 20-feet reflector, though not given in angular quantities, will equally fuffice for the purpofe. By them we have the equatorial diameter to the polar as 103 to 98, or as 1355 to 1289. I have turned the proportion into the latter numbers by way of comparing them the better with the meafures of the 7-feet reflector. By that inftrument the equator of Mars, Oct. 1. we find, was meafured three times; but from the remarks annexed to the different refults, I think the third meafure fhould be ufed. Indeed, on taking the difference of the two firft. which is $34'''$, and dividing by three, we have the quotient $11\frac{1}{3}'''$; then, allotting two-thirds to the firft, becaufe the remark fays pofitively " narrow meafure," it becomes $22'' 34\frac{2}{3}'''$, and taking one-third from the fecond, which is expreffed doubtfully, " rather too full," it becomes $22'' 35\frac{1}{3}'''$: this reflection on the two firft meafures gives additional validity to the third, which is $22''35'''$, or $1355'''$. The polar diameter was meafured twice; and as no reafon appears againft either of the obfervations, I fhall take the mean of both, which is $21'' 29'''$, or $1289'''$; fo that by thefe meafures the equatorial diameter of Mars is to the polar as 1355 to 1289. A lefs perfect agreement between the proportions of the diameters arifing from the meafures of the 20-feet reflector and thofe which we have juft now deduced from the 7-feet, would have been fufficient for our purpofe, as we might eafily have excufed one or two thoufandths of the whole quantity; however, we have no caufe to be difpleafed with this coincidence, though it fhould in part be owing to accident, and therefore fhall admit the above proportion, and proceed to a farther examination of it.

In

In the firſt place, it will be neceſſary to ſee whether any correction be required on account of the different heliocentric and
geocentric ſouth latitude of Mars; which would apparently
compreſs the polar diameter a little, by the defect of illumina-
tion on the north. On computation we find, that a difference
ariſing from that cauſe would give the longitudinal diameter to
the latitudinal as 20000 to 19987; which being much leſs
than one thouſandth part of the whole, may therefore be
neglected.

But next, a very conſiderable correction muſt be admitted,
when we take into account the poſition of the axis of Mars.
The declination of the ſun on that planet, at the time the
meaſures were taken, was not leſs than 27° ſouth; ſo that the
poles were not in the circumference of the diſk by all that
quantity. On a ſuppoſition then, that the figure of Mars is
an elliptical ſpheroid, we are now to find the real quantity of
the polar diameter from the apparent one. It has been proved,
that, in the ellipſis, the exceſſes of any diameters above the
polar one are as the ſquares of the coſines of the latitudes *;
but the diameter at rectangles to the equator of Mars, which
was expoſed to our view in the late oppoſition, was not the
polar one, but ſuch as muſt take place in a latitude of 63°.
Putting therefore $m =$ coſine of 63°, $a = 1355$, $b = 1289$, $x =$
the polar axis, we have $1 : m^2 :: a - x : b - x$. And $\frac{b - m^2 a}{1 - m^2} = x$;
which gives us 1272 nearly, for the polar diameter. The true
proportion, therefore, of the equatorial to the polar diameter
will be as 1355 to 1272; which, reduced to ſmaller but leſs
accurate numbers, is 16 to 15 nearly.

* Aſtr. par M. DE LA LANDE, § 2680.

I ſhall

I fhall now alfo mention fome of the other meafures, but
with a view only to fhew that they are very confiftent with the
above determination. From thofe of the 30th of September,
for inftance, we collect the proportion of the diameters of
Mars as 1340 to 1286; or, reduced to our former numbers,
1355 to 1300. Now, fince thefe meafures were taken the
night before the oppofition, they muft on that account be as
good as the former; and, had thofe of the day of oppofition
not been preferred, becaufe they were oftener repeated, and
the fuperior power of the 7, and great light of the 20-feet
reflector, gave them additional weight, I fhould have taken them
into the account; the very fmall difference, however, cannot
but ftrengthen the refults of the former meafures.

From the obfervations of the 25th of September we have
the proportion of the diameters as 1313 to 1275; and if the
equatorial meafure be increafed in the ratio of 20000 to 19953,
on account of the different heliocentric and geocentric longitude, Mars not being at the full, it will give the ratio of 1316
to 1275; or, conforming to our former numbers, as 1355 to
1312. I have not been very ftrict in the application of the
correction deduced from the phafes of Mars, fince no other
ufe was intended to be made of thefe numbers than merely to
fhew, that they do not very greatly differ from thofe we have
affigned before *.

It

---

* If more ftrictnefs be required, let EC, fig. 32. be the ecliptic; PS its poles;
p s the poles of Mars, and *eq* its equator. Then, the angle *p m* C being found, by
calculation, we fhall have C *m* (radius) to *cm* (cofine of the difference between
the heliocentric and geocentric longitude) as *qv* (fine of the angle *q m v* or *p m* C)
to *ov*. Then, fince with Mars C *c* can never be very great, the fmall triangle
q n o may be taken for fimilar to *qv m*; therefore *q m* (radius) is to *qv* (fine of
p m C)

It was obferved, Oct. 17, 1783, that the equatorial diameter of Mars was ftill greater than the polar, notwithftanding the depredation of the defect of light upon it. On calculating the phafes, we find, that the longitudinal diameter was, that day, to the latitudinal one as 19711 to 20000, which therefore could not be an equal balance to oppofe the fpheroidical figure fo as to render it invifible.

But, Nov. 10. the proportion of the longitudinal diameter to the latitudinal one, from a computation of the phafe of Mars, muft have been as 18762 to 20000 ; and accordingly it was by obfervation found to be more than fufficient to take off all appearance of the polar flattening, and leave a vifible excefs in the axis above the equator.

To obviate any doubts concerning a fallacy that might arife from the convexity of the eye-glafs, or irregular fhape of the fmall fpeculum, I need only refer, for the latter, to the experiments of the 7th and 9th of October, 1783 : for fhould the fhort diameter of my fmall plane fpeculum have occafioned a compreffing of the polar diameter of Mars when expofed to it, half a turn of the telefcope muft bring the other diameter of that fpeculum into the fame fituation, and a contrary effect would have followed. With regard to the former, not only the experiments made with the achromatic, but principally the obfervation with the 20-feet reflector, where I ufed a compound eye-piece magnifying only about 300 times, will fufficiently exculpate the eye-glaffes. It is alfo well known, that in a fingle lens the diftortion of the images, if any fuch there

pmC) as qo ($=qv-vo$) to qn ; which is the required correction or deficiency of the equatorial diameter eq of Mars.

Or, putting $mC = 1$ and $vq = m =$ cofine of the angle Pmp ; it will be $qn = m^2 . cC.$

7   fhould

fhould be, will equally affect the wires of the micrometer, and give a true meafure notwithftanding; and the compound eye-piece I ufed with the 20-feet reflector had likewife the fame advantage, for it is conftructed on the plan lately propofed by Mr. RAMSDEN in the Philofophical Tranfactions *, which he was fo obliging as to communicate to me about a twelve-month ago, and which I immediately adapted to my large micrometers.

On the fubject of the figure of Mars I ought to remark alfo, that perhaps the meafures which were taken of its diameters during the laft oppofition will enable us to afcertain its real fize with greater accuracy than has been done before. The micrometer which can diftinguifh with precifion between the equatorial and polar diameters of this fmall planet, will certainly be admitted as an evidence of confiderable confequence; and fince the refult of thefe meafures is pretty different from what former obfervations give us, I fhould not omit mentioning it.

We have feen that the equatorial diameter, on the day of the oppofition, meafured 22″ 35‴. The diftance of Mars from the earth at that time was .40457, the mean diftance of the earth from the fun being 1; therefore, 22″ 35″ reduced to the fame diftance will be no more than 9″ 8‴.

I fhall conclude this fubject with a confideration relating to the atmofphere of Mars. Dr. SMITH † reports an obfervation of CASSINI's, where " a ftar in the water of Aquarius, at the " diftance of fix minutes from the difk of Mars, became fo " faint before its occultation, that it could not be feen by the " naked eye, nor with a 3-feet telefcope." It is not men-

* Vol. LXXIII. p. 94.
† Optics, § 1096.

tioned

tioned what was the magnitude of the ftar; but, from the circumftance of its becoming invifible to the naked eye, we may conclude, that it muft have been of the fixth or feventh magnitude at leaft. The refult of this obfervation would indicate an atmofphere of fuch an extraordinary extent, fince at the diftance of 36 femi-diameters of the planet it fhould ftill be denfe enough to render fo confiderable a ftar invifible, that it will certainly not be amifs to give an obfervation or two which feem of a very different import.

1783, Oct 26. There are two fmall fixed ftars preceding Mars, of different fizes; with 460 they appear both dufky red, and are pretty unequal; with 278 they appear confiderably unequal. The diftance from Mars of the neareft, which is alfo the largeft, with 227 meafured 3′ 26″ 20‴. Some time after, the fame evening, the diftance was 3′ 8″ 55‴, Mars being retrograde. I faw them both very diftinctly. I viewed the two ftars with a new 20-feet reflector of 18,7 inches aperture, and found them, as I expected, very bright.

Oct. 27. I fee the two fmall ftars again. The fmall one is not quite fo bright in proportion to the large one as it was laft night, being a good deal nearer to Mars, which is now on the fide of the fmall ftar; but when I draw the planet afide, or out of view, I fee it then as well as I did laft night, The diftance of the fmall ftar meafured 2′ 56″ 25‴ *.

* The meafures were accurate enough for the purpofe, though not otherwife to be depended on nearer than, perhaps, fix or eight feconds.

4   The

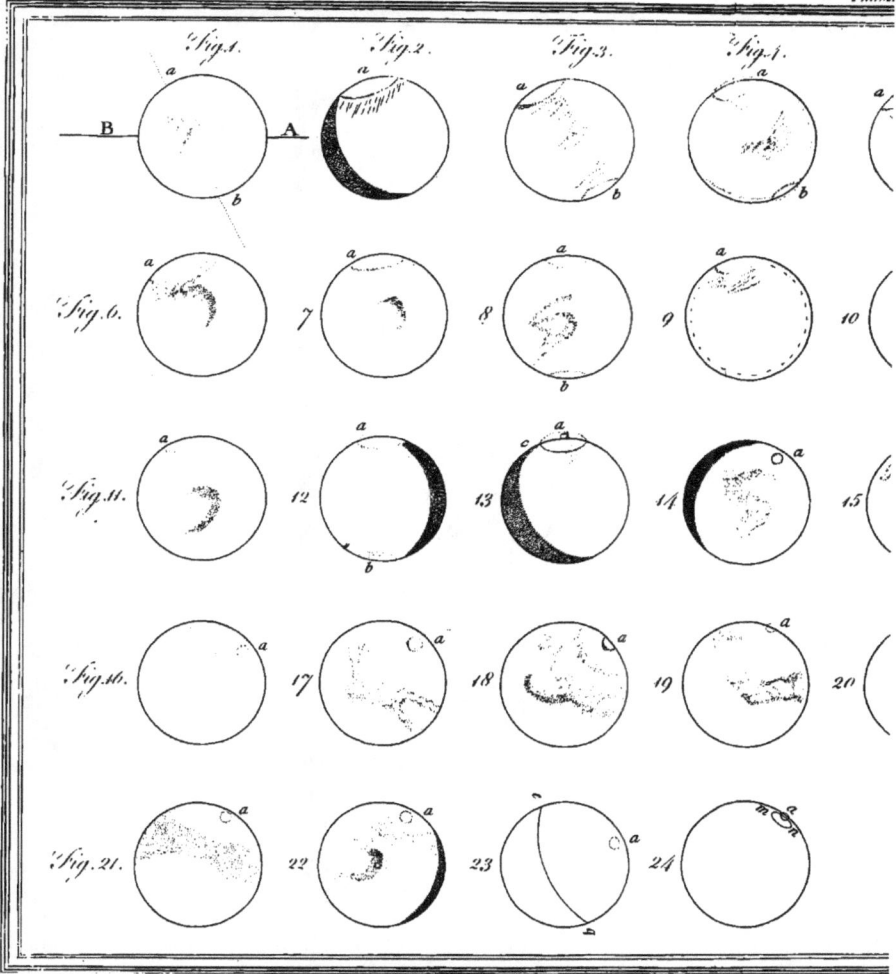

Fig.1. Fig.2. Fig.3. Fig.4.

B ——— A

Fig.6. 7 8 9 10

Fig.11. 12 13 14 15

Fig.16. 17 18 19 20

Fig.21. 22 23 24

Fig. 5.

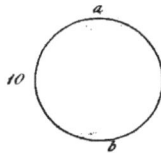

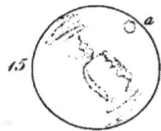

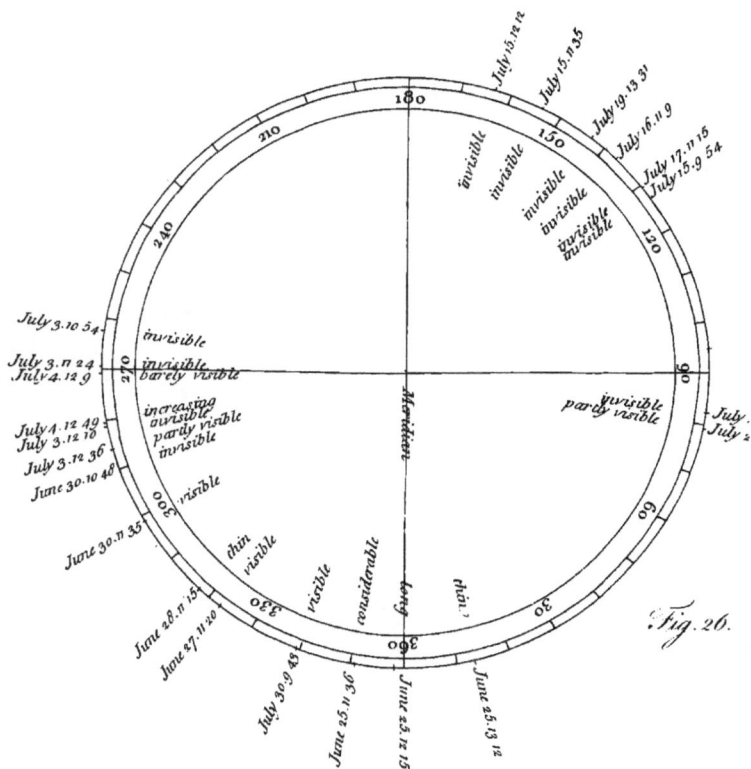

Fig. 26.

*Track of the bright north polar spot on Mars,
in June & July 1781.*

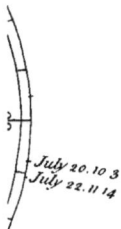

July 20. 10 3
July 22. 11 14

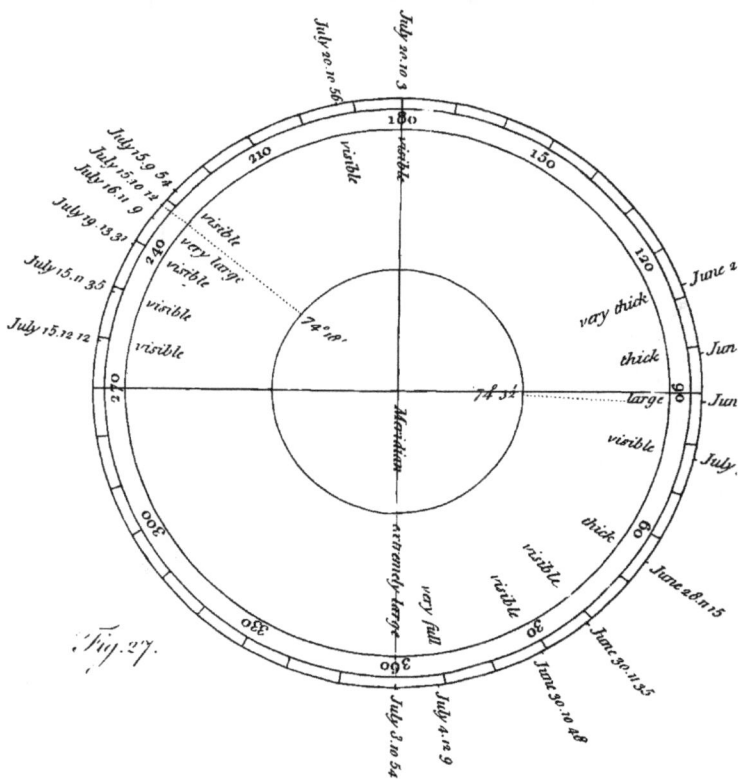

Track of the bright south polar spot on Mars, in June & July 1781.

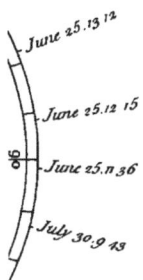

June 25. 13 12

June 25. 12 15

96

June 25. 11 36

July 30. 9 43

e 26 11 15

1,

Track of the bright South polar spot on Mars,
in October 1783.

Philos.Trans.Vol.LXXIV.T

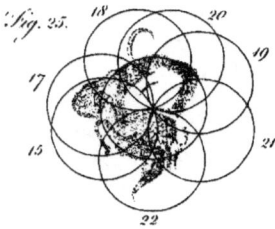

Fig. 25.

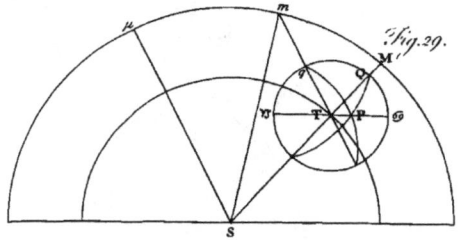

Fig. 29.

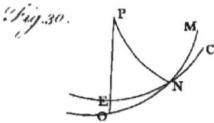

Fig. 30.

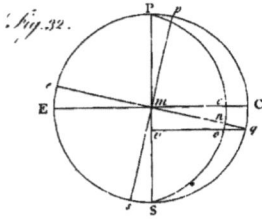

Fig. 32.

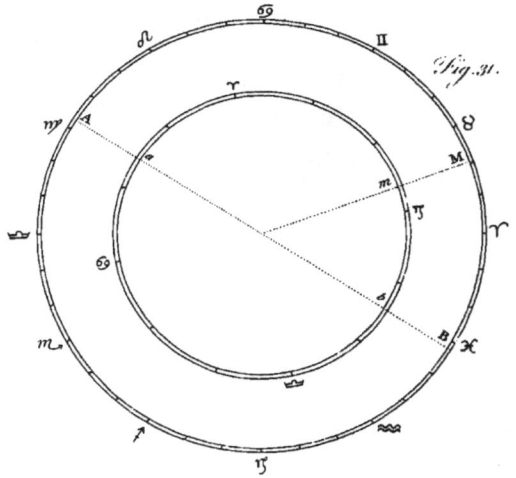

Fig. 31.

The largeft of the two ftars on which the above obfervations were made cannot exceed the twelfth, and the fmalleft the
thirteeenth or fourteenth magnitude ; and I have no reafon to
fuppofe that they were any otherwife affected by the approach
of Mars, than what the brightnefs of its fuperior light may
account for. From other phænomena it appears, however,
that this planet is not without a confiderable atmofphere ;
for, befides the permanent fpots on its furface, I have often
noticed occafional changes of partial bright belts, as in fig. 1.
and 14.; and alfo once a darkifh one, in a pretty high latitude, as in fig. 18. And thefe alterations we can hardly
afcribe to any other caufe than the variable difpofition of clouds
and vapours floating in the atmofphere of that planet.

Refult of the contents of this paper.

The axis of Mars is inclined to the ecliptic 59° 42′.
The node of the axis is in 17° 47′ of Pifces.
The obliquity of the ecliptic on the globe of Mars is 28° 42′.
The point Aries on the martial ecliptic anfwers to our 19° 28′
  of Sagittarius.
The figure of Mars is that of an oblate fpheroid, whofe equatorial diameter is to the polar one as 1355 to 1272, or as
  16 to 15 nearly.
The equatorial diameter of Mars, reduced to the mean diftance
  of the earth from the fun, is 9″ 8‴.
And that planet has a confiderable but moderate atmofphere,
  fo that its inhabitants probably enjoy a fituation in many
  refpects fimilar to ours.

Datchet, Dec. 1, 1783.   W. HERSCHEL.

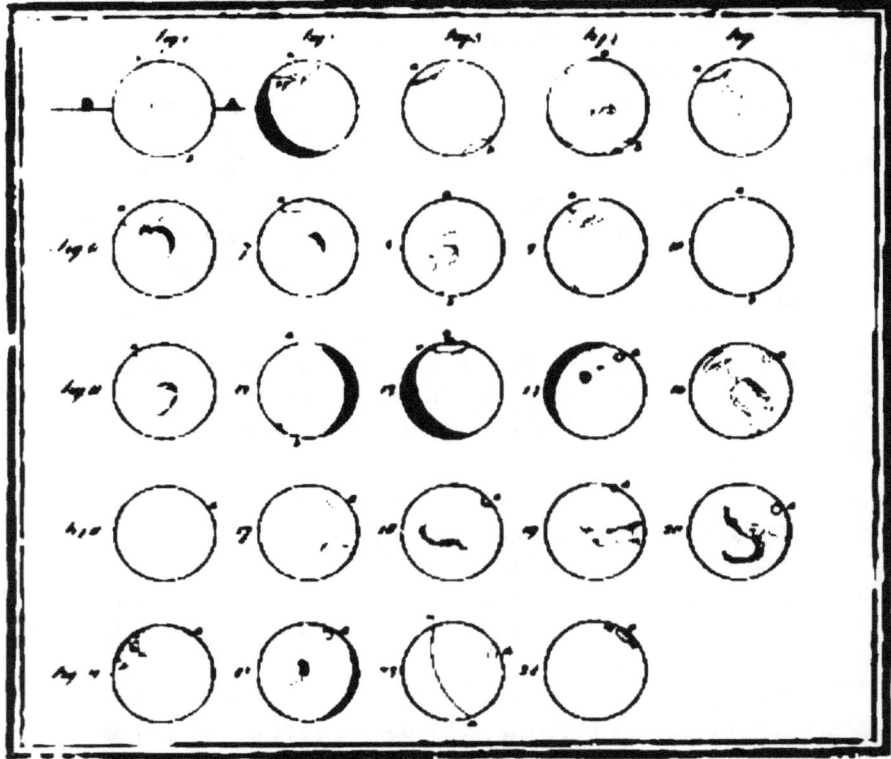

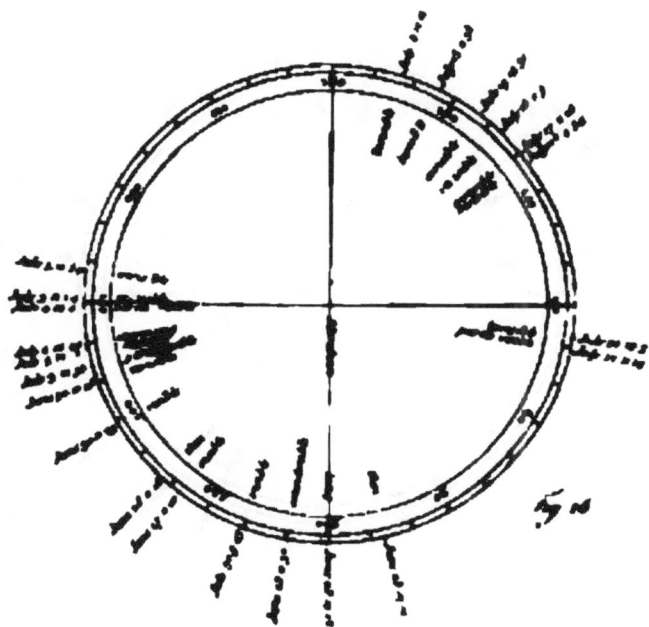

Track of the bright north polar spot as these as seen in July 1711

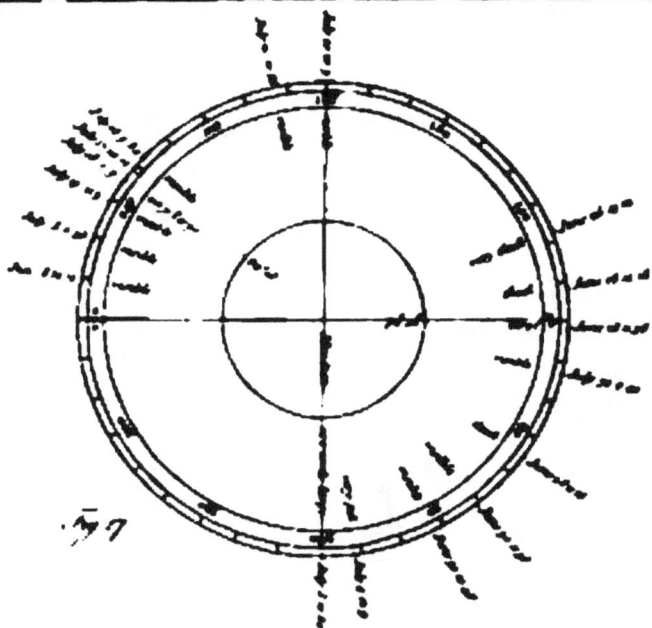

Fig 7

Sketch of the bright south polar spot on Mars
in June & July 1781

www.ingramcontent.com/pod-product-compliance
Lightning Source LLC
Chambersburg PA
CBHW022022190326
41519CB00010B/1572